LIFE SCIENCES CONTRIBUTIONS

ROYAL ONTARIO MUSEUM

NUMBER 104

AF709284

M.B. FENTON

Observations on the Biology of Some Rhodesian Bats, Including a Key to the Chiroptera of Rhodesia

WITHDRAWN

Publication date: 25 September 1975

ISBN: 0-88854-176-7

Suggested citation: Life Sci. Contr., R. Ont. Mus.

QL
1
.T65
no. 104

ROYAL ONTARIO MUSEUM PUBLICATIONS IN LIFE SCIENCES

The Royal Ontario Museum publishes three series in the Life Sciences:

LIFE SCIENCES CONTRIBUTIONS, a numbered series of original scientific publications, including monographic works.

LIFE SCIENCES OCCASIONAL PAPERS, a numbered series of original scientific publications, primarily short and usually of taxonomic significance.

LIFE SCIENCES MISCELLANEOUS PUBLICATIONS, an unnumbered series of publications of varied subject matter and format.

All manuscripts considered for publication are subject to the scrutiny and editorial policies of the Life Sciences Editorial Board, and to review by persons other than Museum staff who are authorities in the particular field involved.

LIFE SCIENCES EDITORIAL BOARD
Chairman: WALTER M. TOVELL
Editor: J. C. BARLOW
Associate Editor: A. R. EMERY
Associate Editor: D. BARR

M.B. FENTON is a Research Associate in the Department of Mammalogy, Royal Ontario Museum, and an Associate Professor of Biology at Carleton University, Ottawa.

PRICE: $1.25
© The Royal Ontario Museum, 1975
100 Queen's Park, Toronto, Canada
PRINTED AND BOUND IN CANADA AT THE JOHN DEYELL COMPANY

Observations on the Biology of Some Rhodesian Bats, Including a Key to the Chiroptera of Rhodesia

Abstract

Weights of 359 bats of 25 species are presented with vocalization frequencies for 14 species of Microchiroptera. Circumstantial evidence of visual orientation while bats were producing echolocation cries was obtained for eight species of microchiropterans. Insect wings obtained under roosts used by *Rhinolophus landeri* (Rhinolophidae) and *Nycteris thebaica* (Nycteridae) indicate differential use of Noctuidae and Sphingidae (Lepidoptera) as food, and selective feeding on particular prey species. Higher levels of flight activity of bats were found near buildings than over adjacent habitats. Frequencies of capture of 254 bats of 20 species in nine habitats at the Hostes Nicolle Institute of Wild Life Research are analysed; brachystegia woodland and areas over large rivers (the Sengwa and the Manyoni) had, respectively, the most and least diverse bat faunas. Availability of roosts affects the bat fauna of an area by selectively excluding species with specific roost requirements, where those requirements are not met. Partitioning of food resources by food type, size of bats, and, for insectivorous forms, wing shape and relative ear size, is discussed. *Laephotis angolensis* (Vespertilionidae) is reported for the first time from Rhodesia. Included here is a key to 61 species of bats known (57) or expected (4) to occur in Rhodesia.

PART I

Introduction

Harrison (1959, 1960, 1962, 1964, 1968) has published several papers on bats which occur in Rhodesia, and recently Smithers (1972) included 52 species of Chiroptera in a checklist of the mammals from Rhodesia. Two additional species have been added to the fauna, *Tadarida bivittata at* Chikupu Caves (17°30′S, 31°20′E; Peterson and Nagorsen, 1975) in May 1972 and subsequently from other parts of the country (Smithers, pers. comm.), and *Laephotis angolensis* which I obtained in January 1974 at the Hostes Nicolle Institute of Wild Life Research (HNIWR; 18°10′S, 28°13′E), in the Sengwa Wild Life Research Area.

Although work on the taxonomy and systematics of African bats has been considerable (summarized by Hayman and Hill, 1971), relatively little has been published on their ecology. A study of the ecology, biology, and systematics of bats in Zaire (then the Belgian Congo) by Vershuren (1957) is an exception, as is the work by Jones (1972) on the ecology of some pteropodids in Rio Muni. Several workers have reported data on reproductive cycles in African bats, for example, Anciaux de Faveaux (1973), Menzies (1973), and Mutère (1973).

The present study was undertaken to gather data on the habitat preferences, activity patterns, and general biology of bats in the vicinity of the HNIWR. Some observations were also made at the Atlantica Ecological Research Station (henceforth Atlantica) near Salisbury (17°53′S, 30°47′E). Most of the data were gathered between 28 December 1973 and 24 January 1974, but some observations from May and June 1972 are also included.

Materials and Methods

Bats were captured using mist nets and a Tuttle Trap (Tuttle, 1974) set over different habitats in the vicinity of the two research stations. At the HNIWR I set some mist nets at 10 m intervals in three lines of 100 m, as well as locating other nets at random in the areas sampled. Bats were weighed on an Ohaus Triple Beam Balance with cage attached, and were banded using Number 2 Bat Bands (4 mm inner diameter). Age was determined by pelage colour and degree of epiphyseal ossification (Davis and Hitchcock, 1965). Voucher specimens were retained for each species and are deposited in the collections of the National Museum of Rhodesia, the Royal Ontario Museum, and the Carleton University Museum of Zoology. Specimens were identified in the field using a key prepared for this study (see Part II).

I monitored activity using an automated ultrasonic sensing system (Fenton et al., (1973) with four ultrasonic sensors tuned to 40 KHz, although on one occasion at Atlantica, a Holgate Ultrasonic Detector tuned to 78 KHz was used on one channel of the system. Results of activity studies from different locations were compared using 't' tests. Additional observations on activity were made using a Zoomar Night Vision Scope. Frequencies of ultrasonic emissions produced by various Microchiroptera as they flew about in a lighted room or verandah were determined using the Holgate Ultrasonic Detector, tunable from 10 to 180 KHz.

Data on the frequencies of capture in different habitats were analyzed using the Shannon-Wiener Index of Diversity ($H' = -\Sigma_{i=1}^{s} p_i \log_e p_i$, where H' is diversity in a group of species, and p_i the relative abundance of the *i*th species measured from 0 to 1.0), or its analogue $B' = -\Sigma_{j=1}^{m} p_j \log_e p_j$ (where B' is habitat breadth of a species, and p_j the relative abundance of the *j*th species measured from 0 to 1.0; Whittaker, 1972). I also used Emlen's (1973) Index of Diversity ($D_v = \Sigma_{i=1}^{s} p_i e^{-p_i}$, where D_v is diversity), and Pielou's (1966) calculation for evenness of samples ($J = H'/H_{max}$, where J is evenness, and H_{max} the $\log_e$ of the number of species caught in the habitat).

Wing shape was determined by the ratio of the lengths of the third to the fifth digit metacarpals (III/V), measurements of length having been obtained with a pair of Helios dial calipers. Relative ear size was obtained by using the ratio of the length of the ear to the length of the forearm (E/FA). The lengths of the ears were obtained using a mm ruler, and the lengths of the forearms with a pair of Helios dial calipers.

Work was conducted in the vicinity of Atlantica where bats were either netted or trapped near the main building, and in the following nine habitats at the HNIWR (the vegetation zones were determined from Cumming (in press) and with the assistance of personnel from the HNIWR):

1. *Colophospernum mopane* woodland, including three distinct areas: (*a*) tall mopane woodland—areas with tall trees in clumps interspersed with a thick growth of grass; (*b*) short mopane woodland—areas of disturbance where the trees were short and there was very little grass; and (*c*) pans (small ponds) in mopane woodland—the area around a series of small pans located in tall mopane woodland.

2. *Brachystegia-Julbernardia* woodland (miombo woodland), henceforth referred to as brachystegia woodland.

3. *Commophora-Combretum* wooded bushland thicket, henceforth referred to as thicket.

4. River fringing woodland and brushland with a diverse flora which in places approaches a forest form, henceforth referred to as riparian forest.

5. Vleis (wet, grassy meadows) surrounded by either mopane woodland (5*a*) or brachystegia woodland (5*b*).

6. Rivers, specifically the Sengwa, the Manyoni, and the Kove, which were either at least 5 m wide (= larger, Sengwa and Manyoni), or less than 2 m wide (= smaller, Kove).

Discussion

Weight

The importance of weight as a factor in faunal structure (e.g., McNab, 1971) and the absence of such data for most species of bats justify their presentation (Table 1). Small sample sizes for all but four of the species reported in Table 1 generally preclude detailed analysis of the data. For some species (*Epomophorus wahlbergi* and *Scotophilus leucogaster*) an effect of age on weight is evident, but this is lacking in other species (*Pipistrellus nanus* and *Eptesicus capensis*).

Table 1. Weights of bats from Rhodesia. Values are means or individual weights.

Species	n	Weights in g			
		Adults		Juveniles	
		males	females[1]	males	females
Epomophorus crypturus	2	—	85.9(2)	—	—
E. gambianus	3	—	100.9(2)	—	71.9(1)
E. wahlbergi†	8	107.7(1)	80.4(2)	—	58.5(5)
*Taphozous mauritianus**	4	—	27.2(4)	—	—
*Nycteris thebaica**	21	10.9(9)	11.2(4)	10.2(6)	10.5(2)
*Rhinolophus denti**	1	7.2(1)	—	—	—
R. hildebrandti	4	24.1(2)	—	20.7(1)	28.6(1)
R. fumigatus	3	13.7(1)	—	—	11.4(2)
R. clivosus	1	—	—	12.2(1)	—
R. landeri	1	6.2(1)	—	—	—
Hipposideros caffer	14	9.2(3)	8.0(3)	7.3(6)	7.2(1)
H. commersoni	7	131.6(3)	—	68.3(3)	53.1(1)
Myotis welwitschii	1	—	14.4(1)	—	—
Nycticeius schlieffeni	36	5.0(2)	5.9(8)	4.7(7)	4.2(19)
Pipistrellus nanus	14	3.3(3)	3.7(7)	3.1(3)	3.5(3)
P. rusticus	3	3.4(1)	4.1(2)	—	—
P. kuhlii	9	3.3(7)	4.0(3)	3.4(1)	3.6(1)
*Eptesicus hottentotus**	1	—	—	14.3(1)	—
E. capensis†	52	5.2(6)	5.9(28)	5.1(7)	5.8(11)
Laephotis angolensis	7	—	7.7(2)	6.0(4)	6.8(1)
Scotophilus nigrita†	6	23.1(1)	23.4(1)	25.2(2)	24.6(2)
S. leucogaster	139	19.0(11)	19.3(44)	15.9(47)	16.2(37)
*Miniopterus schreibersi**	8	10.7(3)	9.5(1)	8.8(2)	8.0(2)
Tadarida nigeriae	13	17.8(11)	19.5(1)	—	15.4(1)
T. bivittata	1	—	—	15.4(1)	—
TOTAL	359				

[1] Post-lactating.
† Taken at Atlantica and HNIWR.
* Taken only at Atlantica.

Prey Selection

I obtained 177 insect wings from the ground inside a hollow baobab tree (*Adansonia digitata*) used as a roost by *Rhinolophus landeri.* These wings represented at least 66 individuals, 92.4 per cent of which were Lepidoptera, the remainder Orthoptera. With the exception of one butterfly (*Charaxes varanes,* 1.5 per cent of the 66 individuals), all of the Lepidoptera were Noctuidae (Catochalinae). *Anua tirhaca* comprised 59.1 per cent of the individuals, whereas *Ophisma lienardi* and *Sphingomorpha cholorea* each accounted for 12.1 per cent. The remainder of the moths included *Achaea illustrata* (4.6 per cent), and an undetermined *Ophisma* sp. (3.0 per cent).

Between 13 November 1969 and 12 March 1970, Rudyerd Boulton collected 389 insect wings from a night roost used by *Nycteris thebaica* at Atlantica. Orthopteran remains accounted for 54 per cent of the insects he obtained, and Lepidoptera 45 per cent. Although I was not able to obtain identifications for the Orthoptera, the Lepidoptera sample included at least 29 species of moths, one of which accounted for 32 per cent of the 87 individual moths present; the next most common species accounted for 3.5 per cent. The most common moth was a sphingid (*Polytychus compar*), and individuals of this family comprised 35.6 per cent of the recognizable moths, whereas noctuids represented 33.3 per cent of this portion. One moth (*Sphingomorpha cholorea*) taken at Atlantica was also present in the baobab sample.

Two of these sphingids (*Hippotion eson* and *Hippotion celerio*) were also present in a collection of insect wings taken from beneath a roost used by *Taphozous perforatus* in Ethiopia (Hill and Morris, 1971), but in neither their sample nor my own did they account for a large portion of the prey selected. Moreover, Hill and Morris (1971) found that 88.9 per cent of their sample was Sphingidae (the remainder Orthoptera), and that one species, *Agrius convolvuli,* accounted for 68.6 per cent of the 51 moths.

Seasonal and geographic variation in insect abundance and the fact that several bats may use a roost make somewhat difficult strict assessment of diet based on insect wings taken from beneath roosts. In spite of these sources of error, the insectivorous bats mentioned above appear to use differently the available insect resources, at least at the familial level. Furthermore, different species of bats appear to feed selectively on some species of insects at certain times. Black (1972, 1974) showed how various species of insectivorous bats in the southwestern United States make differential use of Lepidoptera and Coleoptera as food, and the above data show how several groups of Lepidoptera may account for different portions of the diet of a "moth strategist" (*sensu* Black, 1974).

The most effective demonstration of selective feeding by bats (Buchler, 1973) showed how *Myotis lucifugus* strongly selected mayflies (Ephemeroptera) over other available insects under natural conditions. The high proportions of individual species of Lepidoptera in the samples discussed above may be taken as further evidence of taxon-specific feeding by insectivorous bats.

Echolocation Frequencies and Vision

Frequencies of vocalizations emitted by 14 species of bats flying in a lighted room or verandah (Table 2) indicate that at least *Taphozous mauritianus, Myotis welwitschii, Nycticeius schlieffeni, Laephotis angolensis, Eptesicus capensis, Scotophilus nigrita,* and *Tadarida nigeriae* emit high-intensity sounds that include 40 KHz, and are thus detected by the activity-monitoring equipment. The frequencies I found for *Rhinolophus fumigatus* (50–60 KHz) agree

Table 2. Frequencies of sounds used by echolocating bats.

Species	n	Frequencies in KHz
Emballonuridae		
Taphozous mauritianus	2	30–50
Nycteridae		
Nycteris thebaica	6	70–95*
Rhinolophidae		
Rhinolophus denti	1	70–95
R. hildebrandti	2	45–55
R. fumigatus	2	50–60
Vespertilionidae		
Myotis welwitschii	1	30–75
Pipistrellus nanus	4	70–100
P. kuhlii	4	45–85
Nycticeius schlieffeni	6	40–75
Laephotis angolensis	3	35–65
Eptesicus capensis	6	35–70
Scotophilus nigrita	1	30–60
Miniopterus schreibersi	3	50–80
Molossidae		
Tadarida nigeriae	3	25–80†

* Low intensity sound.
† Accompanied by a continuous, audible rattle.

with Pye and Roberts (1970) who also indicated that *Hipposideros commersoni* and *H. caffer* use 58 to 68 KHz and 140 to 150 KHz respectively, and therefore would not be detected on an apparatus tuned to 40 KHz. Novick (1958) reported that *Pipistrellus ceylonicus* and *P. cormandra* had low and high frequency vocalizations respectively, similar to those observed for *P. kuhlii* and *P. nanus* (Table 2).

In the course of determining the frequencies of echolocation sounds produced by these bats (Table 2), I made the following observations about the use of vision. Upon take-off, all of the individual *T. mauritianus*, *R. denti*, *R. hildebrandti*, *R. fumigatus*, *N. schlieffeni*, *E. capensis*, and *N. thebaica* tested emitted ultrasonic cries as they flew about in the room or verandah, and did not bump into any obstacles. However, after from one to five minutes the aforementioned bats flew directly into glassed or screened windows or doors which they had previously avoided, suggesting a switch in orientation cues (cf. Davis and Barbour, 1965). These bats did not appear to cease production of ultrasonic vocalizations and failed to increase their rates of repetition as they flew toward the door or window. Use of vision was particularly apparent for *N. thebaica*, which flew not only into glassed doors and windows, but also into cupboards, whose glass fronts reflected the outside surroundings.

Nycteris thebaica was the only species that produced only low-intensity sounds and the Holgate microphone had to be kept within 6 cm of the mouth of the flying bat to detect any vocalizations. At Atlantica activity of this species was not detected by the automatic ultrasonic sensing system when operated in the breezeway which was used as a night roost by *N. thebaica*, in spite of considerable activity of this species as observed through the Night Vision Scope.

Activity Patterns

High-intensity echolocating bats (40 KHz) were significantly more active along the front of a bungalow at the HNIWR ($\overline{X}$ = 679 ± 185.3 bat passes, n = 8 nights) than at any of the other sites I studied ($P < 0.005$), and significantly least active along a hallway at the HNIWR headquarters ($\overline{X}$ = 121.5 ± 78.8 bat passes, n = 6 nights, $P < 0.005$). There were no significant differences among

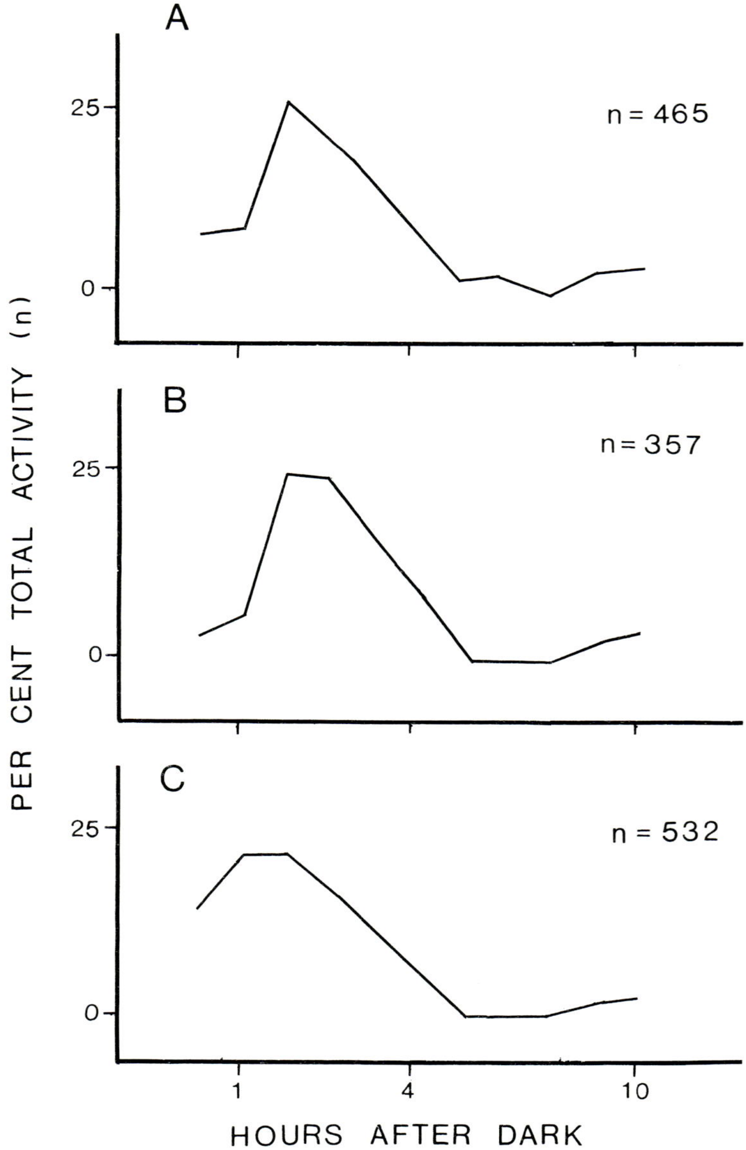

Fig. 1 Activities of bats at the bungalow and adjacent sites (n = number of bat passes).
A. Across the gully.
B. Along the gully.
C. In front of the bungalow.

activity levels at the other sites studied, which included the area across a gully ($\overline{X}$ = 238.4 ± 161.8 bat passes, n = 7 nights) and along the same gully ($\overline{X}$ = 206.7 ± 124.0 bat passes, n = 7 nights) near the bungalow; over the courtyard ($\overline{X}$ = 341.6 ± 53.6 bat passes, n = 5 nights); and near the rim of the escarpment at the HNIWR headquarters ($\overline{X}$ = 355.2 ± 175.1 bat passes, n = 6 nights).

The reasons for the greater activity along the front of the bungalow are not obvious, but could reflect the local distribution of insects, since the lights of the bungalow were on each night until 2300 or 2400 hours and they did attract some insects. Furthermore, the levels of bat activity declined markedly from approximately 2300 hrs, indicating that the highest levels of bat activity corresponded to the period when the lights were on (Fig. 1)

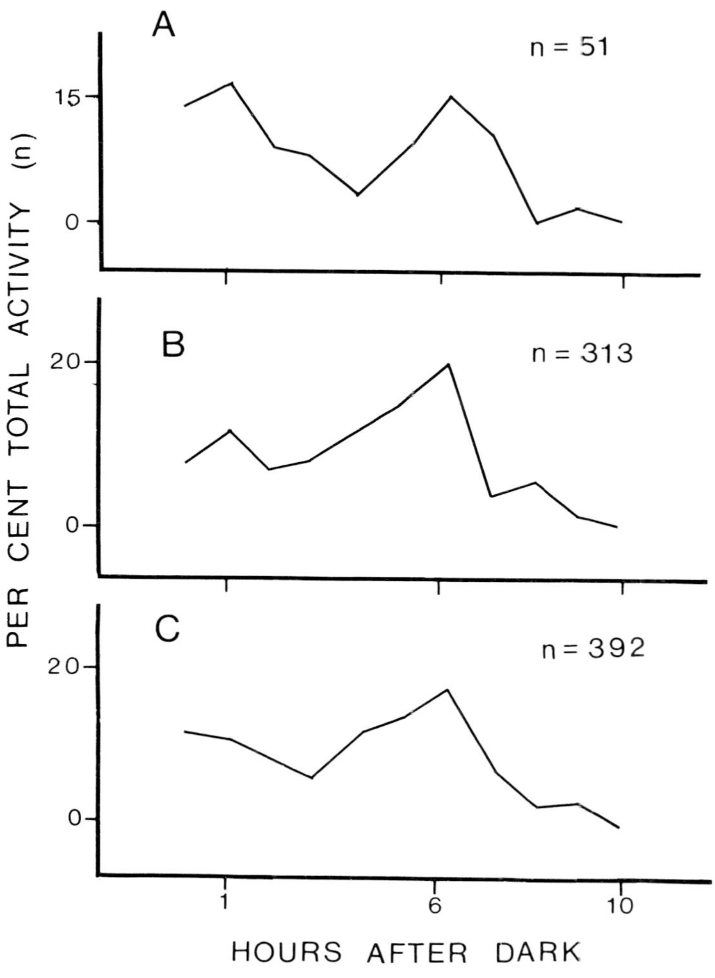

Fig. 2 Activity of bats in the vicinity of the HNIWR headquarters (n = number of bat passes).
A. In the hallway.
B. In the courtyard.
C. Along the rim of the escarpment in front of the headquarters.

Table 3. Distribution of species by habitat.

Species	Mopane woodland: pan	Mopane woodland: tall	Mopane woodland: short	brachystegia woodland	riparian forest	thicket	Vleis: brachystegia	Vleis: mopane	Rivers: larger	Rivers: smaller	n	*B′
Epomophorus wahlbergi	—	—	25	—	50	—	—	—	—	25	4	1.04
E. crypturus	—	—	33	33	—	—	33	—	—	—	3	1.10
E. gambianus	—	—	33	—	33	—	33	—	—	—	3	1.10
Rhinolophus hildebrandti	—	—	—	50	—	50	—	—	—	—	2	0.69
R. fumigatus	—	50	—	—	50	—	—	—	—	—	2	0.69
R. clivosus	—	—	—	—	—	—	—	100	—	—	1	0.00
Hipposideros caffer	—	—	—	100	—	—	—	—	—	—	2	0.00
Hipposideros commersoni	20	40	20	20	—	—	—	—	—	—	5	1.33
Myotis welwitschii	—	100	—	—	—	—	—	—	—	—	1	0.00
Nycticeius schlieffeni	13	16	26	29	6.4	—	3	—	3	3	31	1.72
Pipistrellus nanus	—	100	—	—	—	—	—	—	—	—	1	0.00
P. kuhlii	—	43	—	28	14	14	—	—	—	—	7	1.27
P. rusticus	—	—	—	33	67	—	—	—	—	—	3	0.63
Eptesicus capensis	10	29	5	21	21	2.3	—	—	—	12	42	1.74
Laephotis angolensis	—	83	—	—	17	—	—	—	—	—	6	0.46
Scotophilus nigrita	—	75	—	—	—	—	—	—	25	—	4	0.56
S. leucogaster	21	46	0.7	8	10	—	3	—	11	—	134	1.50
Tadarida nigeriae	—	—	8	—	—	—	8	—	8	78	13	0.80
T. bivittata	—	—	—	—	—	—	—	—	—	100	1	0.00
TOTAL BATS CAUGHT	37	95	16	37	32	3	6	3	18	17	264	
PER CENT OF TOTAL	14	36	6	14	12	1	2	1	7	6		

Column heading: Frequency of bat species by % of n.

$*B' = -\sum_{j=1}^{m} p_j \log_e p_j$

Patterns of activity at the bungalow and adjacent gully sites were the same during all but one of the seven nights studied (Fig. 1), although the number of bat passes varied considerably (see standard deviations above). Therefore the differences in activity were strictly quantitative, suggesting that the bats were not using the bungalow as a night roost. This was confirmed by observation with the Night Vision Scope.

Similarly, levels of activity between the hallway and adjacent courtyard and escarpment rim were different (Fig. 2). Thus bats appeared to use the hallway only as a flyway and moved directly through it, spending more time flying in the less confined courtyard and escarpment areas and accordingly raising the levels of activity there.

Eptesicus capensis, N. schlieffeni, and *L. angolensis*, which were detected by the activity equipment, and *R. hildebrandti, H. caffer*, and *P. nanus*, which were not, were active in the vicinity of the hallway and courtyard *Hipposideros caffer, E. capensis, N. schlieffeni, S. leucogaster, S. nigrita, P. rusticus*, and *P. kuhlii* were active around the bungalow.

Habitat Associations of Bat Species

Distributions of bat species based on capture in mist nets set in the habitats sampled at the HNIWR (Table 3) have been analysed in two ways. First the proportions that each species comprised of the total catches obtained in the different habitats were used to calculate B′ as an indicator of how widespread each of the species was (Table 3). Then the proportions of different species comprising the total catch for individual habitats were used to measure H′ and D_v as indicators of areas with most and least diverse bat faunas (Table 4).

While some species were encountered in only a few habitats, others were more widespread (Table 3). For example, considering the four species for

Table 4. Diversity of bats in different habitats.

Habitat	n_n	n_{sp}	n_i	H′	Dv	J
Mopane woodland						
tall	36	10	95	1.45	0.58	0.57
short	8	8	16	1.65	0.67	0.63
pans	24	4	37	0.79	0.58	0.57
Brachystegia woodland	50	9	37	1.75	0.81	0.80
Thicket	12	3	3	1.10	0.72	1.00
Riparian forest	21	9	32	1.68	0.78	0.76
Rivers						
larger	31	4	18	0.63	0.52	0.46
smaller	12	5	17	1.15	0.69	0.72
Vleis						
mopane woodland	4	3	3	1.10	0.72	1.00
brachystegia woodland	12	3	6	0.87	0.63	0.54

n_n = number of net nights (one spread mist net for one night equals one net night); n_{sp} = number of species; n_i = number of individuals; $H' = -\sum_{i=1}^{s} p_i \log_e p_i$; $Dv = \sum_{i=1}^{s} p_i e^{-p_i}$; $J = H'/H_{max}$

which more than 10 individuals were taken: more than 45 per cent of the *Scotophilus leucogaster* and *Tadarida nigeriae* were captured in one habitat (B′ = 1.50 and 0.80, respectively), whereas *Eptesicus capensis* and *Nycticeius schlieffeni* were encountered in several habitats (B′ = 1.74 and 1.72, respectively).

Brachystegia woodland had the most diverse bat fauna, whereas the least diverse bat faunas were found above large rivers (Table 4). Both diversity indices ranked the above habitats the same, but the other habitats were ranked differently by the two indices (Table 4). Evenness of the sample (J) affects and is affected by the diversity (Table 4); sites with the least even samples have the lowest H′ values, while sites with more even samples have higher H′ values. Exceptions are the vlei in mopane woodland and the thicket, where the samples were small and even (three individuals, three species).

At three of the four sites where data were available for two or more consecutive nights, H′ declines steadily throughout the sampling period (Fig. 3), which possibly may reflect the sensitivity of bats to disturbance (Stebbings, 1969; Fenton, 1970) and the differential use of habitats by the bats. A comparison of the data for *E. capensis*, *N. schlieffeni*, and *S. Leucogaster* will serve to illustrate differences in uses that bats make of different habitats.

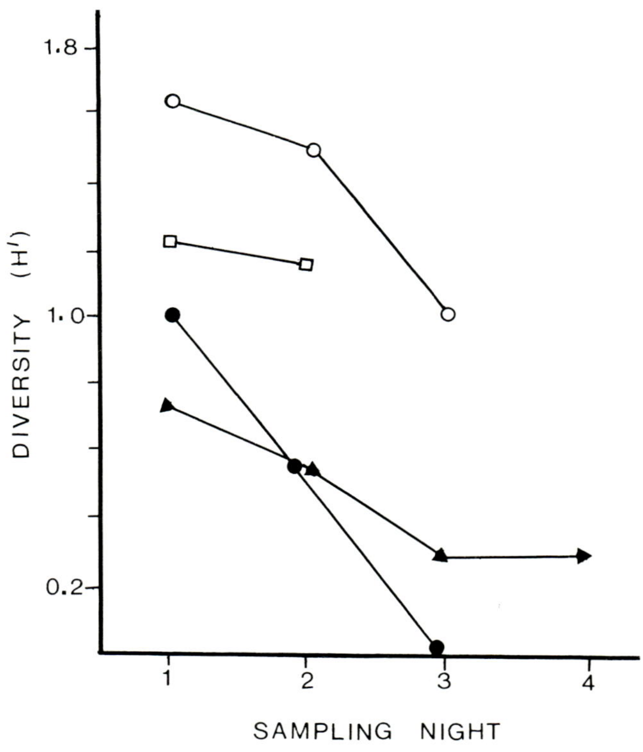

Fig. 3 Change in diversity (H′) throughout the sampling period at four sites: O pan in mopane woodland, □ tall mopane woodland, ● riparian forest, and ▲ hallway at HNIWR headquarters. Data for the first three sites obtained with mist nets, at fourth site with Tuttle Trap.

Nycticeius schlieffeni and *E. capensis* were consistently encountered in four or five habitats respectively (i.e., at least 10 per cent of the catches for each of these species were obtained in four or five habitats; Table 3). In spite of systematic and random netting in the various habitats, none of the 88 marked animals was recaptured. Most netting sites yielded one to three individuals of each of these species, although sometimes five to seven were caught. These data suggest that individuals of both of these species are solitary or live in small groups and are evenly distributed throughout suitable habitat. Since the same distribution patterns were observed for *L. angolensis*, *P. kuhlii*, and *P. rusticus*, albeit based on smaller sample sizes, I suspect that these species have similar roosting habits.

Scotophilus leucogaster, however, was very common in two habitats, less common in three, and rare in two (Table 3). With the exceptions of the areas over the large rivers, this species was most common in tall mopane woodland. Between 30 and 60 minutes after dark, *S. leucogaster* emerged from trees in the tall mopane woodland. Since 92 per cent of the 134 individuals caught were adult females or young, *S. leucogaster* may form nursery colonies in trees. Furthermore, the ratio of adult females to young (1:1.9) suggests that in the study area *S. leucogaster* has twins (44 adult females, 47 juvenile males, 37 juvenile females). The recovery of two banded individuals originally captured over the pans in the mopane woodland, 600 m E at the Sengwa River, suggests that these bats may move to flood plains, perhaps to feed. This is further supported by the fact that less than 4 per cent of the *S. leucogaster* which were taken had fed immediately prior to capture (i.e., the stomachs of the majority were not distended), and over 90 per cent of them were taken just after dark (1900 to 1945 hrs). These observations agree with those summarized for the genus *Scotophilus* by Brosset (1966).

Thus, while some species of bats appear to be solitary or to form small colonies and have relatively even distributions in suitable habitats, others are refuging species (*sensu* Hamilton and Watt, 1970) which form large colonies that are unevenly distributed in an area. Refuging species may roost in one habitat and use other habitats as access routes to feeding areas, since the roosts are often not located in the foraging areas. *Eptesicus capensis* and *N. schlieffeni* appear to be bats of the first category, and *S. leucogaster* a species of the second.

The roost resources of an area can strongly influence the bat fauna there (Humphrey, 1975), since the distribution of refuging or solitary species with stringent roost requirements will be more discontinuous than those species more flexible in their roosting habits. In any region the basic bat fauna will consist of opportunistic species which avail themselves of general roost resources, and an assortment of other species with stricter roost requirements governed by the availability of appropriate roost sites. For example, collecting at the Nuanetsi (21°25′S, 30°45′E) and Humani (20°25′S, 32°07′E) ranches in May and June 1972 produced *Epomophorus crypturus*, *N. schlieffeni*, *E. capensis*, *P. kuhlii*, *Glauconycteris variegata*, and *Tadarida aegyptiaca* at most locations, while other species were only encountered under specific conditions such as around buildings (*Miniopterus schreibersi* and *Tadarida pumila*), baobabs (*Nycteris thebaica*), or kopjes (*Sauromys petrophilus*).

Faunal Structure

Sympatric species of bats may partition food resources by behavioural means (Kunz, 1973), or on the basis of food type and particle size (McNab, 1971; Wilson, 1973; Black, 1974; this study). Various morphological means have been used to demonstrate partitioning of food resources by bats, including dentition (Tamsitt, 1967; Krzanowski, 1971) or wing and ear proportions (Fenton, 1972). In the following discussion, I will consider the bat fauna of the HNIWR on the basis of food type, particle size (bat size), and, for the insectivorous species, wing and ear proportions.

Although Wilson (1973) reported bats from the Ethiopian region occupying several trophic roles, the bats of the HNIWR will be considered as either insectivorous or frugivorous and nectarivorous (some *Epomophorus* spp. appear to feed as much on nectar as on fruit; Rosevear, 1965). The three sympatric species of fruit and nectar feeders at the HNIWR (Table 1) are of two basic sizes, the larger *Epomophorus gambianus* (100 g) and the smaller *E. crypturus* (85 g) and *E. wahlbergi* (80 g). *Epomophorus gambianus* has a more western distribution than either of the other two species (its presence at HNIWR constitutes a range extension from Victoria Falls; Smithers, pers. comm.), and therefore throughout much of Rhodesia only two species of *Epomophorus* are sympatric. Nothing is known about the interactions of these two species. Other fruit-eaters from Rhodesia include the large *Eidolon helvum* (over 200 g), *Rousettus aegyptiacus* (100–130 g) and *R. angolensis* (60–75 g).

The insectivorous species taken at HNIWR (and including *Taphozous mauritianus* which was heard but not captured there) also show different size groups based on weight (Table 1): 16.7 per cent, less than 5 g; 27.8 per cent, 5 to 10 g; 33.3 per cent, 10 to 20 g; 16.7 per cent, 20 to 30 g; and 6 per cent, over 30 g.

I previously demonstrated the value of wing shape (ratio of third to fifth digit metacarpals—III/V) and relative ear size (ratio of length of ear to forearm—E/FA) as indicators of structure in insectivorous bat faunas (Fenton, 1972). When these data are plotted along with weight data for the insectivorous species which are sympatric at the HNIWR (Fig. 4), it is evident that few species are identical in these characteristics. The one *Taphozous* and the two *Tadarida* have longer and narrower wings than any of the other species, but differ from one another in weight and relative ear size. The remaining bats show a spectrum of weight and ear size ranging from the small *Pipistrellus* to the large *Hipposideros commersoni*, or from *Hipposideros caffer* with small ears, to *Laephotis angolensis* with much larger ears. If the size of the bat (weight), its flight characteristics (wing shape), and the nature of its echolocation (relative ear size) affect the prey that is selected, then the data plotted in Fig. 4 are evidence of differential use of food resources by insectivorous Rhodesian bats.

While the data on size (McNab, 1971; or above), as well as information on teeth (Tamsitt, 1967; Krzanowski, 1971), or on wing and ear proportions (Fenton, 1972; or above) appear to indicate partitioning of food by bats, we lack much information about the details. The demonstrated differential use of insects by different bats (Black, 1972; 1974; or above) and selective feeding by some bats (Buchler, 1973) help to support the inferences about faunal structure.

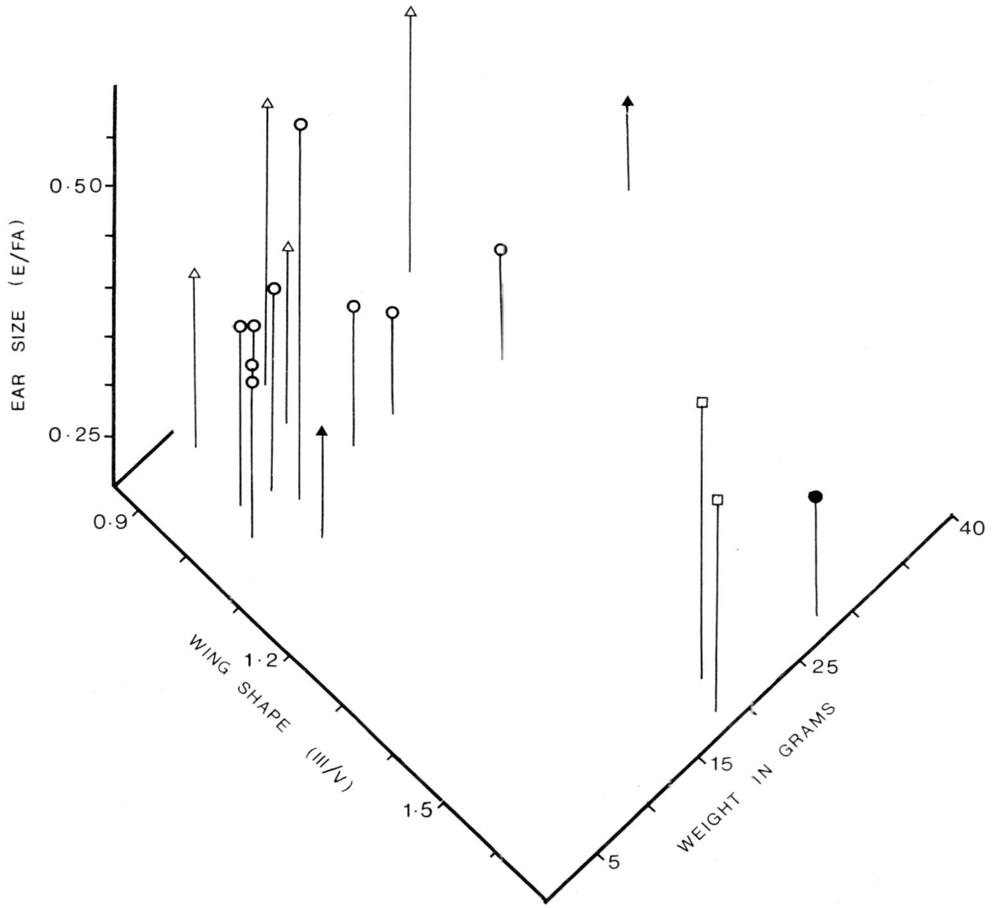

Fig. 4 Structure of the insectivorous bat fauna of the vicinity of the ENIWR as indicated by weight, wing shape, and relative ear size: ● Emballonuridae, △ Rhinolophidae, ▲ Hipposideridae, ○ Vespertilionidae, and □ Molossidae.

PART II

A Key to the Chiroptera of Rhodesia

The necessity of identifying living bats in the field for ecological studies, particularly those involving mark and recapture, prompted the preparation of this key. The key, designed for use with a mm ruler and hand lens, is modified from the work of Hayman and Hill (1971), specific publications on various species (eg., Peterson, 1974; Setzer, 1971; or Peterson and Harrison, 1970), and examination of preserved and living specimens. Sixty-one species are separated out in the key, four of which (marked *) have not yet been reported from the country, but may be expected to occur there. All criteria used in the key may be distinguished on a living specimen in the hand, and because of this the key may be of less value to persons trying to identify skins and skulls or fluid-preserved specimens having closed mouths. Authorities and dental formulae are given as each genus and species is keyed out. The following abbreviations are used: FA = forearm, and E = ear. All measurements are in mm.

In the context of the key, I have resolved several nomenclatural problems as follows: *Taphozous perforatus* includes *Taphozous sudani* (Thomas) and *Taphozous rhodesiae* (Harrison); *Rhinolophus clivosus* is considered separate from *Rhinolophus ferrum-equinum* Schreber as per Koopman (1966); *Tadarida mastersoni* Roberts is treated as a synonym for *T. fulminans* (Hayman and Hill, 1971); and *Tadarida aegyptiaca* includes *Tadarida bocagei* (Seabra).

Other problems of distribution and identity affect several species. Hayman and Hill (1971) reported *Rhinolophus capensis* from Rhodesia, but Smithers (pers. comm.) disputes this record. The specific identity of the *Laephotis* is in question, but I have followed Peterson (pers. comm.) in this matter. Hill (1974) treats *Scotoecus* as a genus (herein considered a subgenus of *Nycticeius*), and shows *Scotoecus hindei* Thomas coming closest to Rhodesia. Although one species of *Miniopterus* is separated in the key, there are probably two species in Rhodesia. The second species, *Miniopterus fraterculus* Thomas and Schwann, is very difficult to distinguish from *M. schreibersi* in the field, especially in the absence of long series, and Hayman and Hill (1971) use skull length to separate them (*M. schreibersi* ca. 15 mm; *M. fraterculus* ca. 14 mm).

I am grateful to Drs. R. L. Peterson and R. H. N. Smithers, who have assisted with the preparation of the key and verified its operation.

Key to Families

1. Claw present on second finger; FA 66–130 .... PTEROPODIDAE, see 6
1′. No claw on second finger; FA usually less than 66 2
  2. Tail fully enclosed by interfemoral membrane 3
    3. Tail terminating in a T-shaped cartilage; ears large; FA 37–66 .... NYCTERIDAE, see 12
    3′. Tail not terminating in a T-shaped cartilage 4
      4. Nose-leaf ornamentation present 5
        5. Nose-leafs arranged as in Fig. 5; FA 37–67 .... RHINOLOPHIDAE, see 16
        5′. Nose-leafs not arranged as in Fig. 5; FA 31–115 .... HIPPOSIDERIDAE, see 25
      4′. Nose-leafs absent; FA 25–65 .... VESPERTILIONIDAE, see 28
  2′. Tail protruding through the interfemoral membrane; FA 58–67 .... EMBALLONURIDAE, see 11
  2″. Tail extending beyond the end of the interfemoral membrane; FA 35–73 .... MOLOSSIDAE, see 45

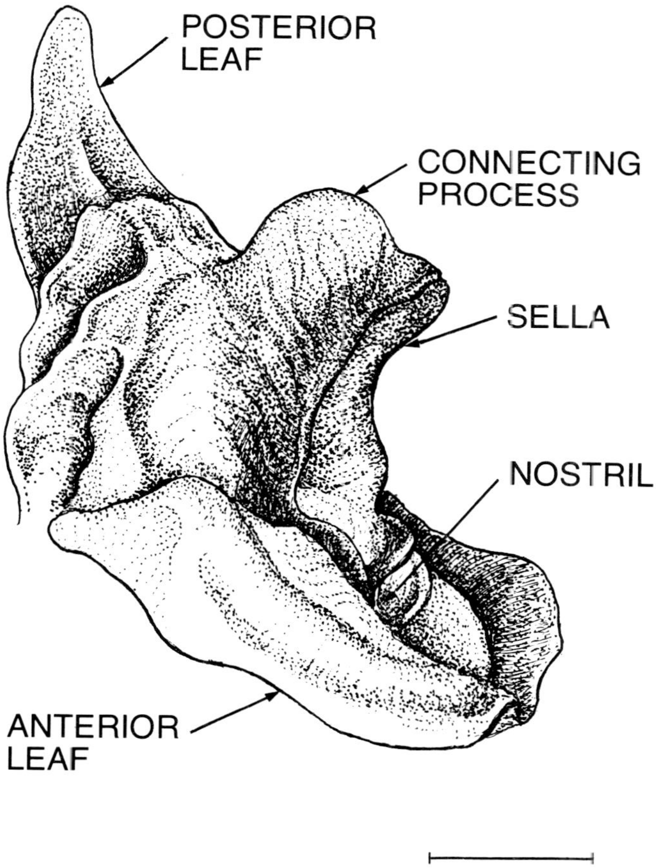

Fig. 5 Nose-leaf of *Rhinolophus clivosus.*

Key to Species

PTEROPODIDAE

6. FA greater than 110; narrow band of fur between wings on dorsum; FA 110–130 .... *Eidolon helvum* (Kerr) $\frac{2\ 1\ 3\ 2}{2\ 1\ 3\ 2}$

6′. FA less than 110 7

7. Ears with basal tufts of white fur 8

8. One post-dental palatal ridge; FA 72–89 .... *Epomophorus wahlbergi* (Sundevall) $\frac{2\ 1\ 2\ 1}{2\ 1\ 3\ 2}$

8′. Two post-dental palatal ridges 9

9. FA males 87–93; females 81–86 .... *Epomophorus gambianus* (Ogilby)

9′. FA males 81–85; females 79–80 .... *Epomophorus crypturus* Peters

7′. Ears without basal tufts of white fur 10

10. Wing membranes insert on first toe; hair short and slick; FA 90–105 .... *Rousettus aegyptiacus* (E. Geoffroy) $\frac{2\ 1\ 3\ 2}{2\ 1\ 3\ 3}$

10′. Wing membranes insert on second toe; hair longer and coarser; FA 66–83 .... *Rousettus angolensis* (Bocage)

EMBALLONURIDAE

11. Fur of dorsum grizzled greyish; venter white; FA 58 -64 .... *Taphozous mauritianus* E. Geoffroy $\frac{1\ 1\ 2\ 3}{2\ 1\ 2\ 3}$

11′. Fur of dorsum bicoloured, not grizzled; FA 60–67 .... *Taphozous perforatus* E. Geoffroy

NYCTERIDAE

12. FA over 55; restricted to closed forest; upper incisors trifid; E 28–35; FA 57–66 .... *Nycteris grandis* Peters $\frac{2\ 1\ 1\ 3}{3\ 1\ 2\ 3}$

12′. FA less than 55 13

13. Upper incisors trifid; E 18–25; FA 36–45 .... *Nycteris hispida* (Schreber)

13′. Upper incisors bifid 14

14. Tragus semilunate (Fig. 6A) 15

15. FA 37–42; E 29–34; colour greyish .... *Nycteris woodi* Andersen

15′. FA 45–50; E 28–34; colour brownish .... *Nycteris macrotis* Dobson

14′. Tragus pyriform (Fig. 6B); E 28–37; FA 42–52 .... *Nycteris thebaica* E. Geoffroy

RHINOLOPHIDAE

16. Face and/or lateral margins of sella (Fig. 5) with long hairs; connecting process low and rounded; greatest breadth of horseshoe usually over 9 .. 17

17. FA 62–67 .... *Rhinolophus hildebrandti* Peters $\frac{1\ 1\ 2\ 3}{2\ 1\ 3\ 3}$

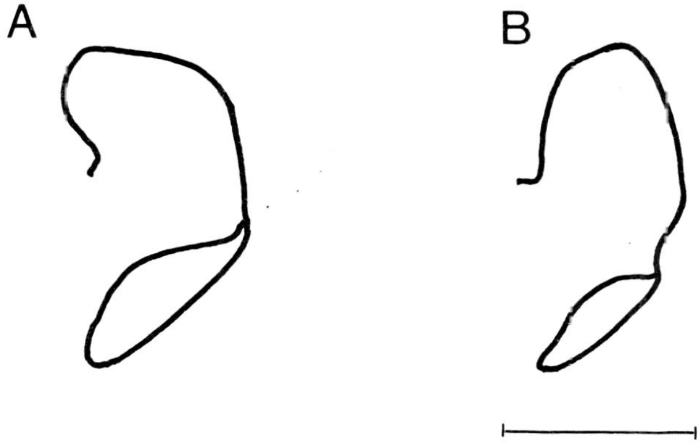

Fig. 6 Tragi of (A) *Nycteris macrotis* and (B) *N. thebaica*.

17′. FA 50–60 .... *Rhinolophus fumigatus* Rüppell

16′. Face and/or lateral margins of sella without hairs; greatest breadth of horseshoe usually less than 9 18

18. Anterior upper premolar external to tooth row; connecting process bluntly pointed; upper canine and p^4 in contact 19

19. FA 50–57 .... *Rhinolophus clivosus* Cretschmar

19′. FA 45–50 .... *Rhinolophus darlingi* Andersen

18′. Anterior upper premolar in tooth row; upper canine and p^4 not in contact 20

20. First phalanx of fourth digit shortened relative to metacarpal; (phalanx less than 8); connecting process to erect point (Fig. 7D); FA 40–48 .... *Rhinolophus landeri* Martin

20′. First phalanx of fourth digit not shortened (over 8) 21

21. Connecting process rises to high, narrow horn (Fig. 7E); FA 44–48 .... *Rhinolophus blasii* Peters

21′. Connecting process low with bluntly pointed tip 22

22. FA 47–51 .... *Rhinolophus capensis* Lichtenstein

22′. FA less than 45 23

23. Ears relatively long; sella broad, venter white; FA 40–46; connecting process as in Fig. 7F .... *Rhinolophus simulator* Andersen

23′. Ears relatively short; sella narrow 24

24. Front edge of connecting process convex; FA 40–46 .... *Rhinolophus swinnyi* Gough

24′. Front edge of connecting process concave (Fig. 7G); FA 37–42 .... *Rhinolophus denti* Thomas

HIPPOSIDERIDAE

25. FA 31–35; nose-leaf small and trident .... *Cloeotis percivali* Thomas $\frac{1\ 1\ 2\ 3}{2\ 1\ 2\ 3}$

25′. FA over 40; nose-leaf not trident 26

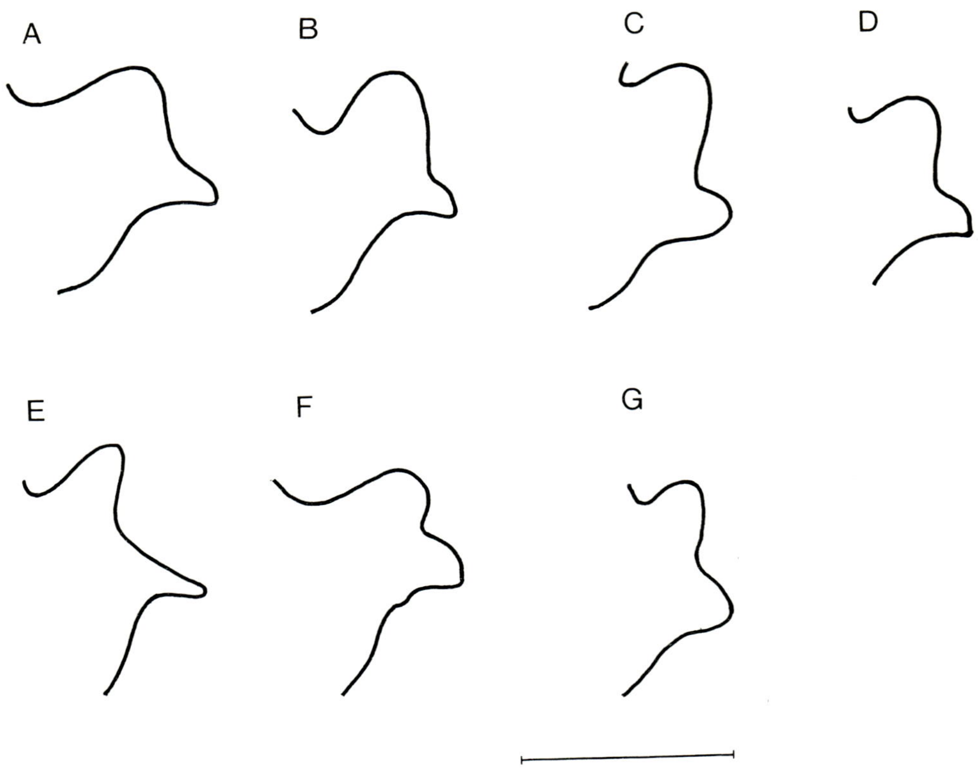

Fig. 7 Connecting processes of (A) *Rhinolophus fumigatus*, (B) *R. clivosus*, (C) *R. darlingi*, (D) *R. landeri*, (E) *R. blasii*, (F) *R. simulator*, and (G) *R. denti*.

26. FA over 75 (79–115) .... *Hipposideros commersoni* (E. Geoffroy) $\frac{1\ 1\ 2\ 3}{2\ 1\ 2\ 3}$

26′. FA 40–55 27

27. FA less than 48; greyish to pale orange fur; main posterior nasal compartment narrow with wide lateral inflations .... *Hipposideros caffer* (Sundevall)

27′. FA over 48; brownish to rufous; main posterior nasal compartment wide with narrow lateral inflations .... *Hipposideros ruber* (Noack)

VESPERTILIONIDAE

28. Second phalanx of third digit equal to three times the first phalanx; FA 42–47 (Miniopterinae) ....*Miniopterus schreibersi* (Kuhl) $\frac{2\ 1\ 2\ 3}{3\ 1\ 3\ 3}$

28′. Ears funnel-shaped with deep emargination below tip; tragus long and narrow with sharply pointed tip; FA 30–39 (Kerivoulinae) 29

29. FA 34–39; venter whitish or buffy .... *Kerivoula argentata* Tomes $\frac{2\ 1\ 3\ 3}{3\ 1\ 3\ 3}$

29′. FA 30–32; venter brown .... *Kerivoula harrisoni* Thomas

28″. Not as above (Vespertilioninae) 30

30. Elongated muzzle; p $\frac{3\text{-}3}{3\text{-}3}$ M$\frac{3\text{-}3}{3\text{-}3}$; fur short and standing away from body; calcar long 31

31. FA 56–57; membranes and FA particoloured (black and red) .... *Myotis welwitschii* (Gray) $\frac{2\ 1\ 3\ 3}{3\ 1\ 3\ 3}$

31′. FA 47–52; membranes not particoloured .... *Myotis tricolor* (Temminck)

31″. FA 36–40; membranes not particoloured; orange-rufous dorsum .... *Myotis bocagei* (Peters)

30′. Muzzle short (less than three premolars) 32

32. Upper incisors 2-2 33

33. Upper premolars 2-2; tragus short and blunt 34

34. Venter pure white; fur extending slightly on to wings; fur of dorsum bicolour (dark base, light tips); membranes pale; FA 34 .... *Pipistrellus rueppelii* (Fischer) $\frac{2\ 1\ 2\ 3}{3\ 1\ 2\ 3}$

34′. Fur on venter dark or light grey; membranes dark 35

35. Tragus hatchet-shaped (Fig. 8A); fur on dorsum bicoloured; FA 25–32 .... *Pipistrellus nanus* (Peters)

35′. Tragus not hatchet-shaped (Fig. 8B) 36

36. Pelage bicoloured above and below; outer upper incisor less than one half inner upper incisor; large upper premolar in contact with canine; FA 30–33 .... *Pipistrellus kuhlii* (Natterer)

36′. Outer and inner upper incisors equal in size; colour more rufous; white border on posterior part of wing membranes between feet and fifth digit well marked; FA 27–30 .... *Pipistrellus rusticus* (Tomes)

33′. Upper premolars 1-1 37

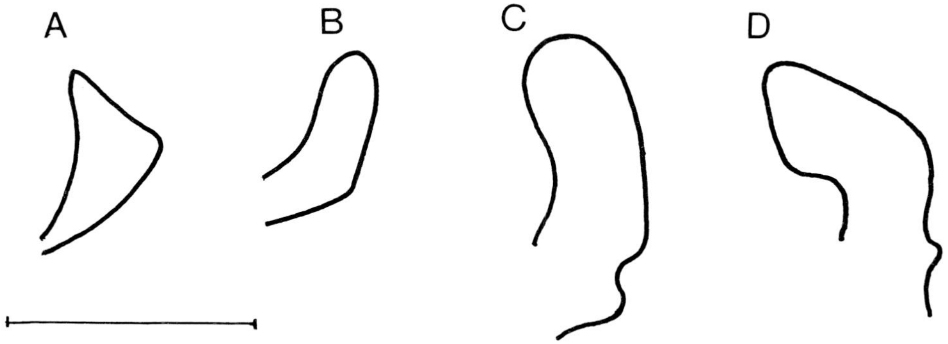

Fig. 8 Tragi of (A) *Pipistrellus nanus*, (B) *P. kuhlii*, (C) *Nycticeius hirundo*, and (D) *N. schlieffeni*.

37. Second phalanx of third digit greater than first; lower lip with lobe at posterior angle 38

38. Wing and interfemoral membrane variegated; fur on venter white and unicoloured; FA 40–45 .... *Glauconycteris variegata* (Tomes) $\frac{2\ 1\ 1\ 3}{3\ 1\ 2\ 3}$

38′. Wing and interfemoral membrane not variegated; fur black at base, narrow white middle band, sandy tips; FA 39–44 .... *Glauconycteris argentata** (Dobson)

37′. Second phalanx of third digit less than first; lower lip without lobe 39

39. Ears over 18 and ca. 50% of FA; FA 36–38 .... *Laephotis angolensis* Monard $\frac{2\ 1\ 1\ 3}{3\ 1\ 2\ 3}$

39′. Ears less than 18; 30–40% of FA 40

40. FA 46–53; membranes dark; inner upper incisor unicuspid .... *Eptesicus hottentotus* (A. Smith) $\frac{2\ 1\ 1\ 3}{3\ 1\ 2\ 3}$

40′. FA less than 45 41

41. Membranes light or translucent; buffy brown dorsum; dirty white venter; FA 34–48 .... *Eptesicus rendalli** (Thomas)

41′. Membranes dark; FA 29–36 .... *Eptesicus capensis* (A. Smith)

32′. Upper incisors 1-1 42

42. FA 40–80; tragus long and tapering 43

43. FA over 70 (70–80); venter white to orange brown .... *Scotophilus gigas* Dobson $\frac{1\ 1\ 1\ 3}{3\ 1\ 2\ 3}$

43′. FA 50–65; venter yellowish .... *Scotophilus nigrita* (Schreber)

43″. FA 43–50; venter beige.. *Scotophilus leucogaster* (Cretzschmar)

42′. FA less than 40; tragus short 44

44. Tragus short and rounded; upper canine grooved on anterior face; cheek teeth $\frac{2\text{-}3}{2\text{-}2}$; penis very elongate; tragus as in Fig. 8C; FA 31–33 .... *Nycticeius hirundo** (de Winton) $\frac{1\ 1\ 2\ 3}{3\ 1\ 2\ 2}$

44′. Tragus as in Fig. 8D; upper canine not grooved on anterior face; cheek teeth not as in preceding; penis not very elongated; FA 30–33 .... *Nycticeius schlieffeni* (Peters) $\frac{1\ 1\ 1\ 3}{3\ 1\ 2\ 3}$

MOLOSSIDAE

45. Ears very large (38–40), joined on projecting snout; collar of pale fur separating darker anterior and posterior parts; FA 62–73 .... *Otomops martiensseni* (Matschie) $\frac{1\ 1\ 2\ 3}{2\ 1\ 2\ 3}$

45′. Not combining above characteristics 46

46. FA over 55 47

47. Ears joined at base; FA 58–66 .... *Tadarida midas* (Sundevall) $\frac{1\ 1\ 2\ 3}{2\ 1\ 2\ 3}$

47′. Ears separate at base 48

48. Venter white; wings translucent; ears large (29–30); white spot on mid-dorsum in adults; FA 57–60 .... *Tadarida lobata* (Thomas)

48′. Venter not white; wings not translucent 49

49. FA 60–67; brown and red forms .. *Tadarida africana** (Dobson)

49′. FA 57–60; brown and red forms .... *Tadarida fulminans* (Thomas)

46′. FA 44–55 50

50. Venter dark or perhaps tinged with grey 51

51. Conspicuous ridge of fur across dorsal surface of hips; variable pattern of white flecks or short stripes laterally on crown and perhaps shoulders and flanks; flanks quite bare; FA 44–49 .... *Tadarida bivittata* (Heuglin)

51′. Conspicuous ridge of fur absent; no white flecks 52

52. Ears separate; no white stripe at ventral contact of wings and body 53

53. Chin and throat blackish; lower canines almost touching at bases; FA 44–47 .... *Tadarida ansorgei* (Thomas)

53′. Chin and throat not blackish; lower canines well separated at bases; FA 44–53 .... *Tadarida aegyptiaca* (E. Geoffroy)

52′. Ears joined at bases; white stripes at ventral contact of wings and body; fur on body black; wings translucent; FA 46–50 .... *Tadarida nigeriae* (Thomas)

50′. Venter light-coloured 54

54. Crown darker than back; underside usually mainly white; E 20–25; FA 44–47 .... *Tadarida nivieventer** (Cabrera and Ruxton)

54′. Crown not darker than back; little or no white on venter; E 28–30; FA 45–50 .... *Tadarida condylura* (A. Smith)

46″. FA 35–44 55

55. Ears joined on top of head; variable pattern of white on venter including lateral stripes and central area of body; wings translucent; FA 37–42 .... *Tadarida pumila* (Cretzschmar)

55′. Ears not joined on top of head; no white on venter; wings not translucent; FA 36–42 .... *Sauromys petrophilus* (Roberts) $\frac{1\ 1\ 2\ 3}{2\ 1\ 2\ 3}$

Acknowledgments

I am especially grateful to R. A. Coutts who accompanied me in the field and to the following who assisted with field work: J. A. Dick, M. S. P. Irwin, R. Miller, E. Mulilo, N. Tsindi, G. E. Turner, P. Wright, Simon, Manuel, and Wuluani. I am most grateful to Rudyerd Boulton, Director of the Atlantica Ecological Research Station, who was instrumental in many parts of the study; and to Dr. D. H. M. Cumming, Director of the HNIWR, and others at that station who contributed immeasureably to the study. I also thank the staff of the National Parks and Wild Life Management of Rhodesia for permitting me to work at the HNIWR and to collect specimens there. To Dr. R.H.N. Smithers, Director of National Museums and Monuments of Rhodesia, who was instrumental in arranging the field trip and who assisted with problems of bat distribution and identification, I am very grateful. In this context, and for considerable other assistance, I thank Dr. R. L. Peterson. I am grateful to Dr. E. G. Munroe of the Entomology Research Institute, Canada Department of Agriculture, for identifying the insect wings and to Rudyerd Boulton for permitting me to use material that he had collected.

Drs. E.R. Buchler, D.H.M. Cumming, D.R. Griffin, T.H. Kunz, R.L. Peterson, J.R. Tamsitt, and G.D. Thomas critically examined the manuscript and made helpful suggestions. Ms. Sophie Poray prepared the figures.

This study was supported by the National Research Council of Canada, by the Department of Biology, Carleton University, and by the Department of Mammalogy, Royal Ontario Museum.

Literature Cited

ANCIAUX DE FAVEAUX, M.

1973 Essai de synthèse sur la reproduction de chiroptères d'Afrique (région faunistique Ethiopienne). Period. biol., 75: 195–199.

BLACK, H. L.

1972 Differential exploitation of moths by the bats *Eptesicus fuscus* and *Lasiurus cinereus*. J. Mammal., 53(3): 598–601.

1974 A north temperate bat community: structure and prey population. J. Mammal., 55(1): 138–157.

BROSSET, A.

1966 La biologie des chiroptères. Paris, Masson. 240 pp.

BUCHLER, E. R.

1973 Prey selection by the little brown bat (*Myotis lucifugus*). A paper presented at the Fourth Annual North American Symposium on Bat Research held in New Orleans in November 1973.

CUMMING, D. H. M.

In press A field study of the ecology and behaviour of warthog. Museum Memoir, Trustees of the National Museums of Rhodesia.

DAVIS, W. H. AND R. W. BARBOUR

1965 The use of vision in flight by the bat *Myotis sodalis*. Am. Midl. Nat., 74(2): 497–499.

DAVIS, W. H. AND H. B. HITCHCOCK

1965 Biology and migration of the bat, *Myotis lucifugus*, in New England. J. Mammal., 46(2): 296–313.

EMLEN, J. M.

1973 Ecology: an evolutionary approach. Reading, Mass., Addison-Wesley. 493 pp.

FENTON, M. B.

1970 Population studies of *Myotis lucifugus* (Chiroptera Vespertilionidae) in Ontario. Life Sci. Contr., R. Ont. Mus., 77: 1–34.

1972 The structure of aerial-feeding bat faunas as indicated by ears and wing elements. Can. J. Zool., 50(3): 287–296.

FENTON, M. B., S. L. JACOBSON, AND R. N. STONE

1973 An automated ultrasonic sensing system for monitoring the activity of some bats. Can. J. Zool., 51(2): 291–299.

HAMILTON, W. H. III AND K. E. F. WATT

1970 Refuging. *In* Johnston, R. F., ed. Annual Review of Ecology and Systematics. Vol. 1. Palo Alto, California, Annual Reviews, pp. 263–296.

HARRISON, D. L.

1959 Report on the bats (Chiroptera) in the collection of the National Museum of Southern Rhodesia, Bulawayo. Occ. Pap. Natn. Mus. Sth. Rhod., 3(23B): 217–231.

1960 Notes on some central and east African bats. Durban Mus. Novit., 6(5): 65–78.

1962 On bats collected on the Limpopo River, with the description of a new race of the tomb bat, *Taphozous sudani* Thomas, 1915. I. Systematic notes. Occ. Pap. Natn. Mus. Sth. Rhod., 26B: 755–769.

1964 Notes on some Southern Rhodesian Microchiroptera. Arnoldia (Rhodesia), 1(3): 1–3.

1968 Further notes on African Microchiroptera. Arnoldia (Rhodesia), 3(3): 1–4.

HAYMAN, R. E. AND J. E. HILL
1971 Order Chiroptera. *In* Meester, J. and H. W. Setzer, eds. The mammals of Africa: an identification manual, pt. 2. Washington, D.C., Smithsonian Institution Press, pp. 1–73.

HILL, J. E.
1974 A review of *Scotoecus* Thomas 1901 (Chiroptera: Vespertilionidae). Bull. Br. Mus. Nat. Hist. (Zool.), 24(4): 169–188.

HILL, J. E. AND P. MORRIS
1971 Bats from Ethiopia collected by the Great Abbai Expedition, 1968. Bull. Br. Mus. Nat. Hist. (Zool.), 21(2): 27–49.

HUMPHREY, S. R.
1975 Nursery roosts and community diversity of Nearctic bats. J. Mammal., 56(2): 321–346.

JONES, C.
1972 Comparative ecology of three pteropid bats in Rio Muni, West Africa. J. Zool., 167(3): 353–370.

KOOPMAN, K. F.
1966 Taxonomic and distributional notes on southern African bats. Puku, Occ. Pap. Dept. Game Fish., Zambia, 4: 155–165.

KRZANOWSKI, A.
1971 Niche and species diversity in temperate zone bats (Chiroptera). Acta Zool. Cracov., 16(15): 683–693.

KUNZ, T. H.
1973 Resource utilization: temporal and spatial components of bat activity in central Iowa. J. Mammal., 54(1): 14–32.

MCNAB, B. K.
1971 The structure of tropical bat faunas. Ecology, 52(2): 352–358.

MENZIES, J. I.
1973 A study of leaf-nosed bats (*Hipposideros caffer* and *Rhinolophus landeri*) in a cave in northern Nigeria. J. Mammal., 54(4): 930–945.

MUTÈRE, F. A.
1973 A comparative study of reproduction in two populations of the insectivorous bats, *Otomops martiensseni*, at latitudes 1°5′S and 2°30′S. J. Zool., 171(1): 79–92.

NOVICK, A.
1958 Orientation in paleotropical bats. I. Microchiroptera. J. Exp. Zool., 138: 81–154.

PETERSON, R. L.
1974 Variation in the African bat, *Tadarida lobata*, with notes on habitat and habits (Chiroptera: Molossidae). Life Sci. Occ. Pap., R. Ont. Mus., 24: 1–8.

PETERSON, R. L. AND D. L. HARRISON
1970 The second and third known specimens of the African molossid bat, *Tadarida lobata*. Life Sci. Occ. Pap., R. Ont. Mus., 16: 1–6.

PETERSON, R. L. AND D. NAGORSEN
1975 Chromosomes of fifteen species of bats (Chiroptera) from Kenya and Rhodesia. Occ. Pap., R. Ont. Mus., 27: 1–16.

PIELOU, E. C.
1966 Species-diversity and pattern diversity in the study of ecological succession. J. Theor. Biol., 10: 370–383.

PYE, J. D. AND L. H. ROBERTS
1970 Ear movements in a hipposiderid bat. Nature, 225: 285–286.

ROSEVEAR, D.
1965 The bats of West Africa. London. Printed by Order of the Trustees of the British Museum. 418 pp.

SETZER, H. W.
1971 New bats of the genus *Laephotis* from Africa (Mammalia: Chiroptera). Proc. Biol. Soc. Wash., 84(32): 259–264.

SMITHERS, R. H. N.
1972 Mammals, list of species known to occur in Rhodesia. Salisbury, National Museums of Rhodesia.

STEBBINGS, R. E.
1969 Observer influence on bat behaviour. Lynx (Praha), n.s., 10: 93–100.

TAMSITT, J. R.
1967 Niche and species diversity in Neotropical bats. Nature, 213: 784–786.

TUTTLE, M. D.
1974 An improved trap for bats. J. Mammal., 55(2): 475–477.

VERSHUREN, J.
1957 Ecologie, biologie et systematique des chiroptères. Expor. Parc Natn. Garamba Miss H. de Saeger, 7: 1–473.

WHITTAKER, R. H.
1972 Evolution and measurement of species diversity. Taxon, 21: 213–251.

WILSON, D. E.
1973 Bat faunas: a trophic comparison. Syst. Zool., 22(1): 14–29.